This set includes 2 walkie-talkie units: these are communication devices that operate on mobile radio frequencies. They have 8(Europe)/ 22(US) channels as well as a back-lit LCD screen, enabling you to communicate over several kilometers free-of-charge (up to 5KM in open areas).

Note: Please read the instructions carefully before using the equipment and retain for future reference.

Specific Recommendations

- To avoid the risk of burns, do not use the device if the aerial is damaged in any way.
- Don not use the device in a potentially explosive setting (e.g. Around pumps, on the lower deck of a boat or around a fuel-storage installation or chemical products).
- If traveling in a car or by bike, stop before using the device.
- Switch off the device if on an airplane or in a hospital.
- Never use the device in close proximity to a radio to avoid interference.
- Remove the batteries if the device is not used for an extended period of time. Never mix used batteries with new ones.
- Keep the transmitter and antenna at least 5cm from your face. Direct the antenna upwards and speak normally.
- Clean the device with a damp cloth. Avoid the use of cleaning agents and solvents.
- Do not modify the device in any way. In the event of damage occurring, ensure that the device is checked by a qualified professional.
- The device cannot be used to contact emergency services.

Functions at Glance

BEFORE USE

Removing the belt clip

Before insert batteries into the device, first pull the belt clip latch away from the device.

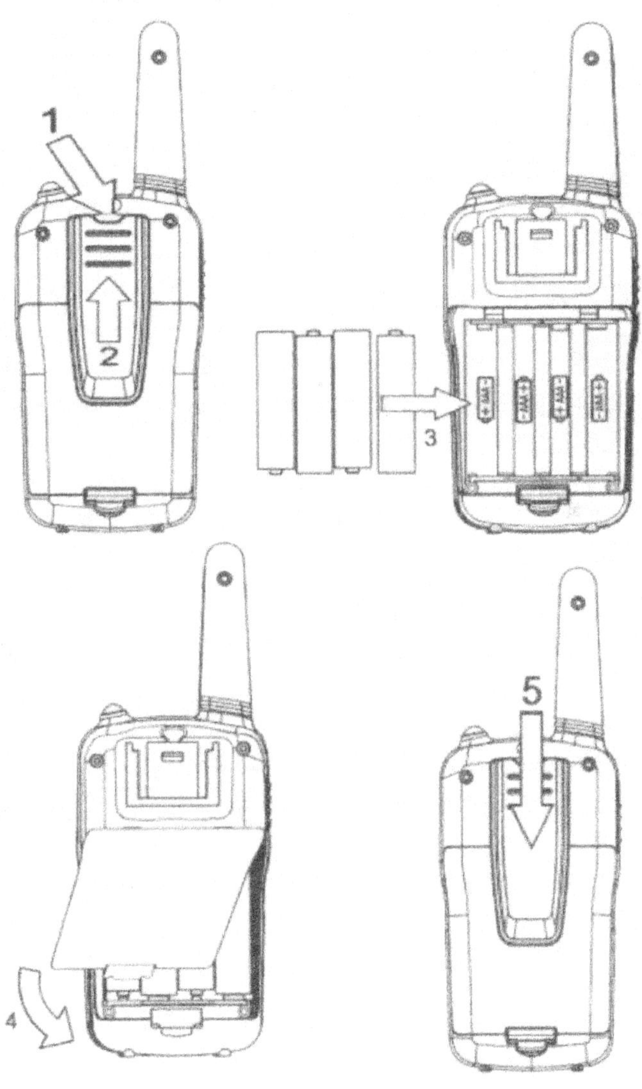

Installing the Batteries

A. FOR USE OF NON-RECHARGEABLE BATTERIES:

1. Slid down the battery compartment cover.

2. Insert 4 x AAA Alkaline Battery (Not included)

3. Position the batteries according to the polarity marking on the battery compartment.

4. After placing batteries into correct positions, put the battery cover back on.

B. FOR USE OF RECHARGEABLE BATTERIES WITH CRADLE AND ADAPTER:

1. If your walkie-talkie comes with a cradle and wall charger, you should only use the battery pack provided to charge. It cannot charge other type rechargeable batteries.

Using the device

1. Turning the device on/off

Switch on the device: Press and hold on the[⏻]button 3 seconds then you will hear a beep sound.

Switch off the device: Press and hold on the [⏻]button 3 seconds then you will hear a beep sound.

2. Adjusting the volume

To increase the volume, press the **Up**[▲]button. And press the **Down**[▼]button to decrease the volume.

Note: VOX mode will be over-ridden when you press the **PTT** button

8. Scanning for an active radio channel

Press and hold the **UP▲** button 2 seconds: The "**SCAN**" function indicator will appear on the display and the channel will scan continuously from 1 to 8 /l to 22. Once an active channel is found, the scanning will stop and you can listen to the transmission.

When the transmission is on the found channel stop, the scanning will resume automatically.

Note: If you press the **PTT** button while listening to a found channel, the device will go back to standby mode on the found channel.

9. Monitor

Press and hold the **DOWN▼** button about 3 seconds to active monitor.

Release the **DOWN ▼** button return to standby mode.

10. Setting the Call tones

The device has 10 call tones.

Press the **MENU** button 4 times, "CA" is displayed and on the current call tone. Press the **UP▲/Down▼** button to change another call tones.

Press the **PTT** button to confirm and return to standby mode.

11. Sending a call tone

Press the PTT button twice time quickly; the call tone will be transmitted on the setting channel.

12. Key-Tone On/Off

When a button is pressed, the unit will beep briefly.

To set the key-tone.

● Press the **MENU**- button (10) five times. "**to**" will be displayed.

● Press ▲ to enable (*ON*) or ▼ disable the Key Tones (*OFF*)

● Press the **PTT**- button to confirm your selection and return to the standby mode.

13. Roger Beep On/Off

After the PTT button is released, the device will send out a Roger beep to confirm that have stopped talking.

Press the **MENU** button six times; "*ro*" will be displayed.

Press the **UP▲/Down▼** button to disable the Roger beep ON/OFF.

Press The **PTT** button to confirm your selection and return to the standby mode.

14. Backlit display

Press any button except for the **MENU** button to activate the backlight of the LCD display. The backlight will light on about 5 seconds.

15. Earpiece connection

The device can be used with an earpiece (If there is an earpiece packed together with the device).

The connector is located on the top of the device.

Insert the earpiece plug into the connector (2.5mm jack).

There is a small "PTT" button on the earpiece that has the same function as the **PTT** button on the device.

When you use the PTT button from the earpiece, you must also use the microphone from the earpiece to talk.

Note: Do not connect other earpieces; it may damage your device.

16. Battery saving function

When the device has not been used for 6 seconds, the economy mode is automatically activated. This does not affect the reception of transmission and the standby mode is automatically re-activated as soon as a signal is detected.

17. Built-in Flashlight

Your device has a built-in flashlight that can be used in sending light signals or for our lighting needs.

18. Lock & Unlock the device

Press and hold the **MENU** button for 3 seconds to lock the device.

Press and hold the **MENU** button for 3 seconds to unlock the device.

19. Technical specifications

Frequency	446.00625MHz~446.9375MHz
	462.5625MHz~467.7125MHz
Channel	8/22 channels
Number Sub-code	CTCSS 99
Transmission Power	$\leq$0.5W
Range	Up to 5 Km in open field
Battery type batteries	4* AAA Alkaline/rechargeable
	(DC 6.0V)
	The battery voltage: 1.5V
Modulation type	FM-F3E
Channel spacing	12.5 kHz (Narrowband)

Channel and Frequency (MHz)
Europe (8CH)

Ch.	Frequency	Ch.	Frequency	Ch.	Frequency
1	446.00625	4	446.04375	7	446.08125
2	446.01875	5	446.05625	8	446.09375
3	446.03125	6	446.06875		

USA (22CH)

Ch.	Frequency	Ch.	Frequency	Ch.	Frequency	Ch.	Frequency
1	462.5625	7	462.7125	13	467.6875	19	462.6500
2	462.5875	8	467.5625	14	467.7125	20	462.6750
3	462.6125	9	467.5875	15	462.5500	21	462.7000
4	462.6375	10	467.6125	16	462.5750	22	462.7250
5	462.6625	11	467.6375	17	462.6000		
6	462.6875	12	467.6625	18	462.6250		

3. Battery Charge Level/Low Battery Indication

The battery charge level is indicated by the number of squares present inside the battery icon on the LCD Screen.

Battery Full

Battery 2/3 charged

Battery 1/3 charged

Battery empty

When the battery charge level is low, the battery icon will flash and a beep will be heard to indicate that the batteries need to be replaced or recharged.

4. Receiving/transmitting communications:

The devices are in "Reception" mode when it is turned ON and no transmitting.

When a signal is received on the active channel, the LCD will display in reception.

When you press the **PTT** (push to talk) button, the device switches to "**Transmission**" mode.

Hold the device in a vertical position with the MIC (microphone) 3-5 cm away from your mouth. While holding the PTT button, speak into the microphone in a normal tone of voice.

Release the **PTT** button when you have finished transmitting.

For others to receive your transmission, they must be on the same channel with you.

Note:

1. The talking range depends on terrain and weather conditions. It will be affected by obstructions such as hills, bush or buildings.

2. Don't try to use two devices which are less than 1.5m (5 feet) apart. Otherwise, you may experience interference.

5. Changing Channels

Press the **MENU** button one time, the channel number will

flash on the display. Press the **up▲ /down▼** button to changethe channel.

Press the PTT button to confirm and return to standby mode.

Note: if no button is pressed within 15 seconds during setting, the device will return to standby mode.

6. CTCSS (Continues Tone Coded Squelch System)

Press the **MENU** button twice, the current CTCSS code will flash on the display. Press the **up▲/down▼** button to change the 99 available codes.

Press the **PTT** button to confirm and return to standby mode.

License-free radio's operating on the 400-470MHZ frequency band, the device has 8/22(option) available radio channels. If there are many device users near you, there is a chance that some of the users are operating on the same radio channel.

When using CTCSS, a low-frequency tone (between 67-250Hz) will be transmitted along with the voice signal. There are 99 available tones to choose from.

You are free to choose one of the 99 available sub-channels. Due to filtering, these sub-channels will generally not be audible so they will not disturb the communication.

7. VOX (Hands free function)

Press the MENU button three times, the current VOX setting will flash on the display and the VOX icon will display,

Press **UP▲** button to set the VOX sensitivity level between 1 and 3 level (level 3 is the high sensitive level).

Press **down▼** button until "*OFF*" appears on the display to turn VOX OFF.

Press the **PTT** button to confirm and return to standby mode.

In VOX mode, the radio will transmit a signal when it is activated by your voice or other sound around you.

VOX Operation is not recommended if you plan to use your device in a noisy or windy environment.

DEFAULT PRIVACY CODES FREQUENCY CHART

Code	Freq (Hz)	Code	Freq (Hz)
1	61.0	2	71.9
3	74.4	4	77.0
5	79.7	6	82.5
7	85.4	8	88.5
9	91.5	10	94.8
11	97.4	12	100.0
13	103.5	14	107.2
15	110.9	16	114.8
17	118.8	18	123.0
19	127.3	20	131.8
21	136.5	22	141.3
23	146.2	24	151.4
25	156.7	26	162.2
27	167.9	28	173.8
29	179.9	30	186.2
31	192.8	32	203.5
33	210.7	34	218.1
35	225.7	36	233.6
37	241.8	38	250.3

Family Radio Service Walkie-Talkie Usage Tips

FRS and GMRS Together

Family Radio Service Usage

<u>License by Rule</u>

The FRS is licensed-by-rule so the general public can use the devices without having to obtain a license. Channel sharing is achieved through a listen-before-talk etiquette. Using FRS does not mean to ignore Part 95 rules.

Channel Sharing

The FRS is authorized 22 channels on the same frequencies as GMRS. All of these are shared with the General Mobile Radio Service. GMRS HT units can talk with FRS units on channels 8-14. Some GMRS mobile and base stations can monitor additional frequencies which could include channels 8-14, which might require programming. According to FCC rules they can not transmit on channels 8-14.

Family Radio Service Limits

Power Limit and Antenna Requirement

FRS Power Limit

FRS units are all HT units.

On channels 8-14 power is limited to 0.5 watt just like GMRS HT units.

When using channels 1-7 and 15-22 FRS units are allowed up to 2 watts.

FRS Antenna Rules

FRS units must have a fixed antenna which can not be removed. This prevents changing its antenna which is prohibited to FRS operators.

If you need a wider area of communications then you should consider GMRS licensed communication.

Visit:

gmrsoperator.com for more information.

FRS Transceivers

Introduction to Transceivers

A FRS transceiver is a two way radio. That means that a transceiver can both transmit a signal to another radio and receive a signal from another radio. The abbreviation for a handheld transceiver is HT.

<u>**Transceiver Classifications**</u>

FRS Transceivers can only be a handheld two-way radio.

About Modulation

Sound Transmission in Air

There are two basic ways to transmit sound over a radio wave in the air. The older technology is amplitude modulation (AM). FRS uses frequency modulation (FM). This should not be confused with FM broadcast radio. A frequency modulation signal rejects radio frequency interference better than an amplitude modulation signal. This is because FM has a larger signal-to-noise ratio than AM. While AM varies the strength of the wave depending upon the audio signal,

FM varies the radio frequency from a center frequency. The maximum variation is the width of the band, which is the frequency deviation.

<u>The FRS UHF Band</u>

GMRS transceivers transmit their signals in the Ultra-High Frequency Band (UHF). The FCC has designated band no. 9 as decimetric waves in the frequency range over 30 MHz up to 300 MHz. FRS is inside this band. MHz is the abbreviation for megahertz. Hertz is a unit designating the number of oscillations per second, which is the frequency.

Visit Author's Web site:

scan QR Code